THE POETRY OF SELENIUM

The Poetry of Selenium

Walter the Educator™

SKB

Silent King Books a WhichHead Imprint

Copyright © 2023 by Walter the Educator™

All rights reserved. No part of this book may be reproduced in any manner whatsoever without written permission except in the case of brief quotations embodied in critical articles and reviews.

First Printing, 2023

Disclaimer
This book is a literary work; poems are not about specific persons, locations, situations, and/or circumstances unless mentioned in a historical context. This book is for entertainment and informational purposes only. The author and publisher offer this information without warranties expressed or implied. No matter the grounds, neither the author nor the publisher will be accountable for any losses, injuries, or other damages caused by the reader's use of this book. The use of this book acknowledges an understanding and acceptance of this disclaimer.

"Earning a degree in chemistry changed my life!"
— Walter the Educator

dedicated to all the chemistry lovers, like myself, across the world

CONTENTS

Dedication v

Why I Created This Book? 1

One - Hope And Care 2

Two - Forevermore 4

Three - For In Selenium 6

Four - Softly Flows 8

Five - Essence Will Glow 10

Six - Science's Grasp 12

Seven - Trace Element 14

Eight - Selenium Divine 16

Nine - Cleanse And Restore 18

Ten - Selenium's Blessings 20

Eleven - Messenger Of Grace 22

Twelve - We Honor Your Might 24

Thirteen - Nature's Maze.	. . . .	26
Fourteen - Selenium's Strength	. . . .	28
Fifteen - Healer, A Protector	. . . .	30
Sixteen - Perfect Reflection	. . . .	32
Seventeen - Oh, Selenium	. . . .	34
Eighteen - Radiant Soul	. . . .	36
Nineteen - Myriad Ways	. . . .	38
Twenty - Every Magnitude	. . . .	40
Twenty-One - Everlasting Lease	. . . . . .	42
Twenty-Two - Awe Of Your Decree	. . . . .	44
Twenty-Three - Turns Into Wonders	. . . . .	46
Twenty-Four - Power Of Selenium	. . . . . .	48
Twenty-Five - Life's Grand Scheme	. . . . .	50
Twenty-Six - Captured Our Hearts	. . . . .	52
Twenty-Seven - Pure And Rare	. . . . . .	54
Twenty-Eight - Wisdom Does Bloom	. . . . .	56
Twenty-Nine - Celestial Grace	. . . .	58
Thirty - Profound Vibration	. . . .	60
Thirty-One - Purges The Toxins	. . . . . .	62
Thirty-Two - Catalytic Abilities	. . . . . .	64

Thirty-Three - Guardian True	. . . .	66
Thirty-Four - We Thank You	. . . .	68
Thirty-Five - Honor Selenium	. . . .	70
About The Author	. . . .	72

WHY I CREATED THIS BOOK?

Creating a poetry book about the chemical element of Selenium was an interesting and unique endeavor. Selenium, with its atomic number 34, has several fascinating properties and applications that can inspire poetic exploration. Poetry has the power to capture the essence of an element, its symbolism, and its impact on the world. By delving into the characteristics of Selenium, such as its role in biology, its use in technology, and its environmental significance, this poetry book can provide a creative and thought-provoking perspective on this element. It can also serve as a means to educate and raise awareness about the importance of Selenium in various contexts.

ONE

HOPE AND CARE

In the depths of Earth, where secrets lie,
A wondrous element, Selenium, does reside.
A shimmering metal, with a touch of grace,
Its story unfolds, in this poetic space.

Born in the stars, forged in cosmic fire,
Selenium emerged, with celestial desire.
A gift from the heavens, a rare cosmic jewel,
With properties unique, it begins to rule.

In nature, it dwells, in soil and sea,
In tiny amounts, yet vital it be.
An element of life, a trace in our breath,
Selenium, the guardian of health.

Within our bodies, it quietly abides,
A protector, a warrior, by our side.

Defending against toxins, it takes its stand,
Selenium, the guardian, lending us a hand.
From the depths of the earth, to the tips of the skies,
Selenium's beauty, never ceases to surprise.
The alchemist's dream, the chemist's delight,
A treasure that shines, in the darkest night.
So let us celebrate this element rare,
Selenium, a symbol of hope and care.
For in its presence, we find strength and grace,
A testament to the wonders of our cosmic space.

TWO

FOREVERMORE

In a realm beyond our mortal sight,
Where stars ignite eternal light,
A wondrous element was born,
From cosmic fire, it was adorned.

Selenium, a treasure of the skies,
With shimmering grace, it mesmerizes,
In nature's tapestry, it weaves,
A guardian of health, it achieves.

Within our bodies, it takes its place,
A trace of hope, a touch of grace,
Protecting cells from harm's cruel hand,
A shield against life's shifting sands.

Oh Selenium, with beauty rare,
A jewel that glimmers in the air,
In darkest night, you softly gleam,
A beacon of hope, a gentle beam.

    From the depths of Earth's embrace,
You rise with grace, with gentle pace,
A symbol of care, a gift so rare,
A reminder of wonders beyond compare.
    So let us cherish your cosmic birth,
And honor you upon this Earth,
For you, Selenium, we shall adore,
Forevermore, forevermore.

THREE

FOR IN SELENIUM

In the realm of celestial fires,
Where stars ignite their cosmic pyres,
A wondrous element was born,
Its essence, like a stellar adorn.
    Selenium, a treasure of the skies,
With powers that mesmerize,
A guardian of health it became,
With properties that bear no shame.
    Bathed in radiant cosmic light,
It brings vitality, pure and bright,
A protector of cells, strong and true,
Selenium, we owe our thanks to you.
    From the heavens, it descended down,
Its purpose to heal, to mend, to crown,

A beacon of hope in the darkest night,
Guiding us with its gentle light.

Through the veins of Earth, it flows,
Where life's energy ebbs and grows,
Unveiling secrets, unraveling strife,
Selenium, the essence of life.

So let us cherish this cosmic gift,
Embrace its power, let our spirits lift,
For in Selenium, we find the care,
That binds us all, beyond compare.

FOUR

SOFTLY FLOWS

In the cosmic dance, a star was born,
A radiant gem, on heavens adorned.
Selenium, the guardian of health,
A beacon of hope, its essence felt.
   From nebula's embrace, it emerged,
With brilliance and grace, it surged.
A celestial gift, bestowed with care,
Selenium, a symbol, beyond compare.
   In the depths of Earth, it found its place,
A mineral treasure, with hidden grace.
Nurtured by nature's gentle hand,
Selenium, a protector, took its stand.
   In fields of green, where life takes hold,
Selenium's touch, like stories told.

A guardian of vitality, it imparts,
Nourishing the body, healing hearts.
    Its power extends beyond the skin,
Defying darkness, letting light in.
A shield against harm, a guardian's might,
Selenium, the defender, shining bright.
    Through rivers and streams, it softly flows,
In every creature, its presence shows.
A whisper of care, a gentle embrace,
Selenium, the healer, brings solace and grace.
    Oh, Selenium, we marvel at your might,
A cosmic wonder, a guiding light.
In nature's tapestry, you play your part,
A precious element, forever in our heart.

FIVE

ESSENCE WILL GLOW

In the depths of the cosmos, a shimmering star,
A celestial desire, born from afar.
Selenium, the guardian of health's grace,
A metal that dances in ethereal space.

Its atomic number, thirty-four it holds,
An element of stories yet to be told.
With silver-white luster, it glimmers afar,
A treasure of the skies, a celestial czar.

Within the human body, it finds its place,
A protector of cells, a beacon of grace.
Defending against harm, it takes its flight,
Shielding the body, shining through the night.

A trace of Selenium, a gift from above,
Nurturing life, spreading care and love.
A guardian of wellness, a symbol so rare,
A cosmic essence, beyond compare.

So let us embrace this heavenly trace,
And cherish the Selenium's healing embrace.
In every molecule, a promise it keeps,
To nurture our bodies, in its care we seek.
   Oh, Selenium, celestial star,
With powers so vast, you've traveled far.
In every atom, your essence will glow,
A guardian of health, forever bestowed.

SIX

SCIENCE'S GRASP

In nature's realm, where wonders lie,
A guardian stands with watchful eye.
Selenium, an element rare and true,
A symbol of health, for me and you.
    Within the soil, it finds its home,
Absorbing light where shadows roam.
A trace of grace, a gift divine,
Its presence, a treasure, so benign.
    From valleys deep to mountaintops high,
Selenium spreads its healing sigh.
A protector, it shields with might,
Against the perils of the darkest night.
    In every cell, it plays its part,
A catalyst for life, a work of art.

With every breath we take, we owe,
To Selenium's care, a debt we show.
    Its power lies not just in science's grasp,
But in the harmony of nature's clasp.
It whispers secrets of the universe vast,
A cosmic beauty, unsurpassed.
    Oh, Selenium, a celestial gem,
In your essence, we find a precious hymn.
A symbol of hope, a beacon of light,
Guiding us through the darkest night.

SEVEN

TRACE ELEMENT

Selenium, guardian of health,
In your presence, we find wealth.
A symbol of hope and care,
With powers beyond compare.

In the depths of the Earth, you reside,
A beauty that cannot be denied.
Your atomic number, thirty-four,
Unveils the secrets you hold in store.

A defender against toxins you are,
Protecting us near and far.
With antioxidant might, you fight,
Keeping our bodies shining bright.

A trace element, so vital and rare,
You're found in the soil, water, and air.
Nourishing life, from plants to beasts,
You bring harmony to nature's feasts.

Oh Selenium, cosmic in your birth,
A gift from the heavens, with infinite worth.
Healing wounds and soothing pain,
You bring solace amidst life's strain.

A beacon of light, guiding us through,
Your presence, a promise, forever true.
Selenium, oh element divine,
In your embrace, we find peace entwined.

EIGHT

SELENIUM DIVINE

In the realm of elements, Selenium resides,
A guardian of health, where harmony presides.
Within its atomic shell, secrets it keeps,
As it shields us from harm, while the world weeps.

A shield against toxins, a warrior so brave,
Selenium protects us, steadfast and grave.
From pollutants it shields, with a resolute might,
An ally in darkness, a beacon of light.

Its presence, a blessing, in every cell,
A healer, a guide, where wonders dwell.
With gentle touch, it mends what's broken,
Guiding us on a path, where wellness is spoken.

In the fields it dances, where nature thrives,
Nourishing the land, where life survives.

Through rivers it flows, a shimmering stream,
Enriching the earth, fulfilling a dream.
   Oh Selenium, guardian of our well-being,
Protector of life, in every living being.
With grace and power, you fulfill your role,
A shelter in chaos, a balm for the soul.
   So let us embrace you, dear Selenium divine,
In your presence, health and harmony align.
For in your element, we find solace and care,
A treasure, a gift, beyond compare.

NINE

CLEANSE AND RESTORE

Selenium, the guardian of wellness,
A cosmic wonder in nature's caress.
In the depths of the earth, you reside,
A precious gift, from heaven derived.
   Your presence brings vitality anew,
Healing souls, making spirits true.
With gentle touch, you nourish the land,
Transforming chaos into harmony so grand.
   Like a beacon of light, you shine so bright,
Guiding us through darkness, day and night.
Protector of life, defender of the weak,
Your power is humble, yet far from meek.
   Oh Selenium, we bow to your might,
As you shield us from toxins, day and night.

A warrior against harm, you stand tall,
A shield for our bodies, protecting us all.
   With antioxidant magic, you cleanse and restore,
Bringing peace and solace, forevermore.
We thank you, Selenium, for your grace,
For being the light in this cosmic space.

TEN

SELENIUM'S BLESSINGS

In the realm of elements, a guardian's born,
A silvery treasure, a healer adorned.
Selenium, the mystic, with powers profound,
A solace it brings, where harmony is found.
　In the body's temple, it takes its place,
A protector of cells, with grace and embrace.
Defending against toxins, it stands tall,
Shielding the fortress, where wellness befalls.
　A cosmic essence, it holds within,
Connecting the stars, where miracles begin.
Radiating energy, pure and bright,
Guiding our souls through the dark of the night.
　A whisper of balance, it softly imparts,
Mending the wounds, healing shattered hearts.

With every breath, it breathes life anew,
Revitalizing spirits, making dreams come true.
    Oh Selenium, the mighty and wise,
A jewel of the earth, that never denies,
The power it possesses, the wonders it weaves,
Bringing vitality, where hope never leaves.
    So let us honor this element divine,
For its gifts are abundant, like stars that align.
In harmony and solace, may we forever dwell,
With Selenium's blessings, our stories we'll tell.

ELEVEN

MESSENGER OF GRACE

In the realm of elements, there lies a gem,
A guardian of wellness, we call it Selenium.
With grace and strength, it protects our core,
A precious presence, forevermore.

Like a shield, it wards off toxins and strife,
A defender of life, in the cosmic dance of life.
From the depths of the earth, it emerges bright,
Guiding us through darkness, with its healing light.

Selenium, oh Selenium, we sing your praise,
For the miracles you bestow in countless ways.
You mend what's broken, with gentle touch,
Bringing solace and healing, we cherish so much.

In the fields of nature, you bring harmony,
Nourishing life with your divine alchemy.

From the smallest creatures to the towering trees,
Your touch brings vitality, to all that breathes.

    Oh Selenium, celestial messenger of grace,
In your presence, we find a sacred space.
With gratitude and admiration, we bow our heads,
For the blessings you bring, as we tread.

    In the tapestry of creation, you hold secrets untold,
A precious element, shining bright, and bold.
Selenium, oh Selenium, we honor your might,
Forever grateful, for your cosmic light.

TWELVE

WE HONOR YOUR MIGHT

In the realm of creation, a hidden gem,
Lies a silver element, Selenium, its name.
A guardian of life, it silently dwells,
In the depths of the Earth, where its power swells.
    With a gentle touch, it heals and mends,
A miraculous force, on which life depends.
Its essence dances in the autumn breeze,
As golden leaves fall from the trees.
    Selenium, the alchemist of the land,
A master of balance, with a steady hand.
It weaves harmony through nature's tapestry,
A symphony of life, in perfect harmony.
    From the smallest seed to the tallest tree,
Selenium's touch, a gift of purity.

It shields and protects, with a loving embrace,
Nourishing life with its grace.

    Oh Selenium, we sing your praise,
For the wonders you bring in nature's maze.
A symbol of hope, a beacon of light,
Guiding us through the darkest night.

    In gratitude, we honor your might,
For the blessings you bring, shining bright.
Selenium, the element so divine,
Forever grateful, for your presence, we pine.

THIRTEEN

NATURE'S MAZE.

In the depths of Earth's embrace,
Where minerals weave their mystic grace,
A spark of cosmic light I find,
Selenium, a treasure of a kind.

Born in the heart of ancient stars,
Your essence shines, no matter how far.
With gentle touch, you heal and mend,
A guardian of life, my cosmic friend.

Through the soil, your tendrils creep,
Awakening nature from her slumber deep.
With every breath, you bring new life,
Banishing darkness, soothing strife.

Protector of the land, defender true,
You shield us from toxins, old and new.
With your alchemy, you transform,
Purifying nature, in every form.

    From the oceans to the skies above,
Your presence radiates, a symbol of love.
Harmony you bring, in every measure,
A symphony of balance, a cosmic treasure.
    Oh Selenium, I sing your praise,
For the miracles you weave in nature's maze.
For nourishing life, both near and far,
I salute you, bright and shining star.

FOURTEEN

SELENIUM'S STRENGTH

In the realm of nature's art, Selenium gleams,
A radiant light, the land's sweetest dreams.
With gentle touch, it nurtures the soil,
A healer, a guardian, a blessing to foil.

Beneath the surface, where secrets reside,
Selenium's magic, it cannot hide.
With grace and poise, it mends the land,
Bringing harmony, where chaos once fanned.

A protector, it stands against the tide,
Defending the earth with unwavering pride.
Through trials and tribulations, it remains,
Selenium's strength, nothing can restrain.

Its presence, a shield, against harmful might,
A valiant defender, day and night.

In the depths of darkness, it finds its way,
Guiding lost souls, to a brighter day.

Oh Selenium, we honor your might,
Shining forth, a beacon of light.
The guardian of balance, the bringer of peace,
Your cosmic essence, forever shall not cease.

So let us sing, a song of praise,
To Selenium's power, in wondrous ways.
For in your presence, the world is blessed,
With vitality, harmony, and eternal zest.

FIFTEEN

HEALER, A PROTECTOR

In realms unseen, where secrets hide,
There lies a metal, beaming with pride.
Selenium, a guardian of light,
Emerges from the depths, gleaming bright.

A master of healing, it does possess,
A touch of grace, to mend and bless.
With powers untold, it soothes the pain,
Restoring harmony, like a gentle rain.

In nature's laboratory, it does reside,
A protector, a shield, by its side.
Against the toxins, it stands tall,
Detoxifying, purifying, for one and all.

But beyond the realm of the physical plane,
Selenium's essence, it does sustain.

A guide for souls, in search of truth,
It connects the spirits, from old to youth.
  With wisdom deep, it opens the mind,
A bridge to realms, where answers we find.
In dreams it whispers, a song of light,
Guiding the lost, through the darkest night.
  Oh Selenium, a metal divine,
A shimmering jewel, in the grand design.
A healer, a protector, a beacon of hope,
In your presence, our spirits elope.
  Let us embrace your gifts so rare,
And bask in your glow, beyond compare.
For in your elements, we find our grace,
Selenium, our guardian, in every space.

SIXTEEN

PERFECT REFLECTION

In the realm of elements, Selenium prevails,
A gift from the universe, its power never fails.
A healer, a mender, it mends the broken,
In the depths of darkness, its light is awoken.

Selenium, the mender, with a touch so divine,
It heals the wounded, brings vitality to mine.
Through its mystical essence, it whispers a tune,
Restoring harmony, like a symphony in the moon.

Through the alchemy of Selenium, we find solace,
A potion of life, a powerful elixir to embrace.
It nourishes the body, like a gentle caress,
Reviving the spirit, banishing distress.

Oh, Selenium, protector of the weak,
A guardian of health, a defender we seek.

Against the toxins that threaten our core,
You stand strong, ensuring we suffer no more.
    Guide us, Selenium, on a path of connection,
Unite us with nature, in perfect reflection.
With your presence, the spirits intertwine,
Creating a symphony, where harmony shines.
    Gratitude we owe, to Selenium's might,
For its gifts of healing, bringing love and light.
Oh, Selenium, we sing your praise,
Forever grateful for your mystical ways.

SEVENTEEN

OH, SELENIUM

In the realm of elements, a mystical hue,
Lies a gem adorned with celestial dew,
Selenium, the alchemist's prized creation,
Unveils its secrets, a symbol of transformation.
    A guardian of balance, it quietly resides,
In the tapestry of life, where harmony abides,
Its touch mends the broken, heals the unwhole,
A potion of mending for the weary soul.
    Within its essence, a cosmic light gleams,
Guiding lost wanderers through their dreams,
A protector of life, a beacon of hope,
Selenium's radiance helps us to cope.
    In the alchemist's hands, it weaves its spell,
Turning leaden sorrows into golden wells,

An elixir of wisdom, a potion divine,
Selenium's alchemy, a treasure benign.
    Oh, Selenium, we sing your praise,
For the gifts you offer, in infinite ways,
Nourishing the Earth with your divine light,
Guiding us towards peace, from darkness to bright.

EIGHTEEN

RADIANT SOUL

In the realm of nature's symphony,
Where harmony and balance meet,
Lies a treasure, rare and serene,
A shimmering dance, the element of Selenium.
    A guardian of purity, it takes its place,
In the hallowed grounds of Earth's embrace,
With a gleaming glow, it illuminates,
The path to wisdom, it navigates.
    Selenium, oh radiant soul,
Protector of the environment's whole,
Through your essence, we find solace,
In the healing embrace of your grace.
    With tender touch, you cleanse the air,
Purifying the world, beyond compare,

A steadfast guardian, a shield of light,
Guiding us through darkened night.
　　Selenium, oh divine guide,
In your presence, we confide,
You bring connection, you bring peace,
A bridge between worlds, a spiritual release.
　　In your sacred embrace, we find,
A union of body, heart, and mind,
For you are more than just an element,
You are a source of enlightenment.
　　Selenium, we sing your praise,
With humble reverence, our voices raise,
For in your essence, we behold,
Nature's secrets, untold.
　　So let us cherish your gentle might,
And bask in your ethereal light,
For in the realm of elements, so rare,
Selenium, you are beyond compare.

NINETEEN

MYRIAD WAYS

In the realm of elements, a gem so rare,
Selenium, a wonder beyond compare.
A guardian of nature, a healer divine,
With powers untold, a mystical shrine.
    Through fields and forests, this element roams,
Bringing new life, as nature's backbone.
It purifies the air, with a touch so light,
Bringing harmony, restoring nature's might.
    In the depths of darkness, it's a guiding light,
Illuminating souls, banishing the night.
A defender of spirits, a beacon of grace,
Selenium's presence, a celestial embrace.
    With strength unmatched, it stands tall and true,
A warrior of elements, a force that grew.

It shields the vulnerable, in times of despair,
With Selenium's might, all darkness, it shall bear.
    Oh, Selenium, we sing your praise,
For the miracles you weave, in myriad ways.
A healer, a protector, a gift from above,
Your essence, a testament to eternal love.

TWENTY

EVERY MAGNITUDE

In the realm of elements, Selenium shines bright,
A protector, a healer, a beacon of light,
With atomic number thirty-four, it resides,
In the periodic table, where its power presides.

A guardian of health, Selenium defends,
Against the ailments that our bodies contend,
A trace of this element, a dose of grace,
It shields our cells, a divine embrace.

Selenium, the mender of broken spirits,
With wisdom it guides, dispelling all limits,
It whispers serenity to troubled minds,
Restoring peace, where tranquility finds.

Through the trials of life, it leads the way,
A compass of hope, where darkness may stray,
Selenium, a defender, unwavering and true,
In its presence, courage and strength renew.

A source of enlightenment, Selenium imparts,
Knowledge and insight, to open our hearts,
It illuminates paths, once shrouded in doubt,
Guiding us forward, as we journey about.

With every breath, Selenium purifies the air,
Removing toxins, with utmost care,
Restoring balance to nature's domain,
A gift from the earth, its benevolent reign.

In the soil, Selenium nourishes and feeds,
Enriching the land, fulfilling its needs,
For all that it offers, we express our gratitude,
Selenium, a treasure, in every magnitude.

TWENTY-ONE

EVERLASTING LEASE

In the realm of elements, Selenium shines,
A guardian of health, a friend of mine.
With powers so potent, it heals and defends,
A marvel of nature, where beauty transcends.
    Selenium, a shield against darkness and strife,
Bringing balance and harmony to life.
Its presence, like a radiant beam,
Guiding us through the realms unseen.
    In alchemical lore, it holds the key,
Transforming base metals with alacrity.
A catalyst for change, a magnum opus,
Selenium, the essence of transformation and focus.
    Its mystical aura, a shimmering glow,
A touch of magic to make hearts aglow.

From the depths of Earth, it does arise,
A gift from the heavens, a treasure in disguise.
    Selenium, the alchemist's delight,
Unveiling secrets hidden from sight.
With its touch, lead turns into gold,
A testament to its power so bold.
    Oh Selenium, your wisdom profound,
A beacon of hope, forever renowned.
You purify air, water, and soul,
Restoring balance, making us whole.
    Grateful we are for your gifts so divine,
Selenium, the compass of hope, ever shine.
In your presence, we find solace and peace,
A testament to your everlasting lease.

TWENTY-TWO

AWE OF YOUR DECREE

In the realm of Selenium, a mystic dance unfolds,
Where healing powers and secrets untold,
Converge in harmony, a celestial embrace,
Guiding souls on paths of wisdom and grace.

From the depths of nature, Selenium emerges,
A guardian of life, as the universe converges,
With its aura of light, it illuminates the way,
A beacon of hope in the darkest of days.

In the soil it dwells, a giver of life,
Nurturing growth, banishing strife,
A catalyst of transformation, it weaves,
A tapestry of existence, where beauty breathes.

Oh Selenium, we sing your praise,
For the wonders you grace upon our days,

Your presence, a blessing, an alchemist's touch,
Transforming the mundane into wonders as such.
From health to vitality, you bestow,
A tonic for body and spirit to grow,
Your healing touch, a gentle embrace,
Revealing paths to wholeness and grace.
Selenium, oh Selenium, we honor your might,
In the cosmic symphony, you shine so bright,
A luminary guide, forever we'll be,
Grateful for your gifts, in awe of your decree.
In Selenium's embrace, we find solace and light,
A mystical dance, forever igniting our sight,
With gratitude and reverence, our souls alight,
In the realm of Selenium, our spirits take flight.

TWENTY-THREE

TURNS INTO WONDERS

In the realm of elements, a steadfast guardian,
Selenium, a presence that never wanes,
With grace it shines, a source of enlightenment,
A bridge between worlds, where dreams remain.

In nature's realm, it forms a sacred bond,
A backbone strong, a defender of spirits,
With alchemical touch, it weaves a magic wand,
A beacon of grace, as the darkness it limits.

Through the ethereal fog, it casts its charm,
A healer it is, with soothing embrace,
Binding wounds, restoring bodies warm,
A protector, a presence, a saving grace.

It purifies the air, the water it cleans,
A unifying force, in harmony it brings,

Guiding us on a journey, where hope redeems,
Selenium, a gift, from celestial strings.
    With gratitude, we honor your mystical aura,
Through your alchemical touch, we find our way,
In your presence, the mundane turns into wonders,
Selenium, a blessing, forever we'll say.

TWENTY-FOUR

POWER OF SELENIUM

In the realm of elements, a gem so rare,
Selenium, a gift beyond compare.
A unifying force, it holds the key,
To heal the world and set our spirits free.

With gentle touch, it mends the broken,
Restoring balance, words unspoken.
A healer of wounds, both deep and wide,
Selenium, the soothing balm inside.

Through darkest nights, a guiding light,
Selenium shines, banishing the blight.
It clears the air, purifying the breath,
Returning nature's might, conquering death.

A protector, strong and steadfast,
Selenium shields us from the tempest's blast.

With mighty grace, it stands its ground,
A guardian, ever watchful, always found.

    In its presence, enlightenment gleams,
Selenium's wisdom, beyond our dreams.
It whispers secrets of the universe's lore,
Unveiling mysteries, forevermore.

    Oh Selenium, we sing your praise,
For the harmony you bring, in myriad ways.
Grateful we are for your precious gift,
A source of light, our spirits you uplift.

    In every atom, every bond we see,
The power of Selenium, set us free.
With gratitude, we honor your name,
And forever, in our hearts, you'll claim.

TWENTY-FIVE

LIFE'S GRAND SCHEME

In a realm where nature's secrets abide,
There lies a mystical element, dignified.
Selenium, a force both pure and rare,
With power to heal, restore, and repair.

   A guardian of balance, it does hold,
Through its touch, harmony unfolds.
In the soil, it nurtures life's grand design,
Ensuring vitality, growth, and divine.

   With grace, it purifies the water's flow,
Removing impurities, letting purity show.
A guiding light in the darkest of night,
Selenium's glow brings wisdom and insight.

   Like a beacon, it illuminates the way,
Leading seekers to truth without delay.

Its presence, a blessing, a source of grace,
Enlightening minds in a celestial embrace.
　　Oh, Selenium, we honor your might,
A symbol of enlightenment shining bright.
In your essence, we find solace and peace,
A catalyst for wonder, where miracles increase.
　　So let us celebrate this celestial gem,
A testament to nature's alchemical emblem.
Selenium, a guardian of life's grand scheme,
In your mystical aura, we find solace and dream.

TWENTY-SIX

CAPTURED OUR HEARTS

In the realm of elements, a mystic light gleams,
A touch of divinity, a dance of dreams.
Selenium, the alchemical wonder, so rare,
A celestial gift, beyond compare.
    Through its atomic dance, a healing aura unfurls,
A balm for the weary, a salve for the hurts.
With gentle grace, it mends the broken,
Restoring life's essence, with words unspoken.
    Selenium, the gentle guardian of air,
A purifier of breath, a solace so fair.
It captures the toxins, the pollutants that roam,
And transforms them into life's sweetest perfume.
    Oh, Selenium, your presence divine,
A beacon of light, a guide through time.

In the darkest of nights, you shimmer and glow,
Leading us forward, when lost, we may go.
　　Through the depths of water, you cleanse and you purify,
A guardian of purity, a savior nearby.
With every ripple, every drop that you touch,
You restore harmony, a gift that means so much.
　　Selenium, the alchemist's treasure,
A symbol of wisdom, a mystical measure.
We honor your magic, your transformative might,
In awe of your power, we bask in your light.
　　So, let us sing praises to Selenium's name,
A guardian, a healer, forever the same.
In gratitude, we embrace your divine art,
For you, dear Selenium, have captured our hearts.

TWENTY-SEVEN

PURE AND RARE

In the realm of alchemy, Selenium gleams,
A mystical element, beyond mere dreams.
With healing powers, it touches the soul,
A guiding light, making broken hearts whole.
    Selenium, the purifier, cleansing the air,
A guardian of purity, beyond compare.
Through its touch, impurities dissolve,
Cleansing the spirit, problems resolve.
    In nature's embrace, Selenium resides,
Amidst the earth's treasures, it silently hides.
Its essence, like moonlight, softly shines,
A gentle glow, a gift divine.
    Through the darkest night, it guides the lost,
A beacon of hope, no matter the cost.

Selenium, the light that leads the way,
Illuminating the path, come what may.

    We give thanks for Selenium's grace,
For its healing touch, in every place.
A mystical element, so pure and rare,
We honor its presence, with love and care.

    So let us embrace Selenium's might,
A symbol of transformation, shining bright.
In its wisdom, we find solace and peace,
Selenium, our guide, our eternal release.

TWENTY-EIGHT

WISDOM DOES BLOOM

In the realm of elements, a gem does reside,
A mystic force, with secrets it hides.
Selenium, the healer, it holds in its core,
A transformative power, forevermore.
    Within its essence, a touch of grace,
A celestial light, that illuminates space.
From the depths of the earth, it emerges strong,
A catalyst of change, where miracles belong.
    Selenium, the protector, with an armor so pure,
Defending the body, ensuring a cure.
With every breath, it shields from harm,
A guardian of health, a mystical charm.
    In its presence, wisdom does bloom,
Unveiling truths, dispelling gloom.

Selenium, the sage, with knowledge profound,
Guiding the lost, where enlightenment is found.
 Through its magic touch, impurities fade,
Purifying the soul, a serenade.
A beacon of purity, a gleaming light,
Selenium, the bringer, shining so bright.
 So let us embrace this element divine,
A force of nature, a treasure to find.
For in Selenium's essence, we shall see,
A world of wonders, waiting to be free.

TWENTY-NINE

CELESTIAL GRACE

In the realm of alchemical touch,
Where mystic forces blend and clutch,
Lies a shimmering element, pure and bright,
Selenium, guardian of ethereal light.

Through the alchemist's ancient art,
Selenium weaves its healing part,
A protector of life, a shield from harm,
Guiding us through chaos and alarm.

With gentle grace, it cleanses the air,
A mystical presence beyond compare,
Purifying waters, bringing harmony's tide,
Selenium's essence, a tranquil guide.

In the depths of darkness, it sparks a flame,
A beacon of enlightenment, no two the same,

Whispering secrets of the universe untold,
Selenium's wisdom, a treasure to behold.
  In the tapestry of creation, it finds its place,
A guardian, a warrior, with celestial grace,
Oh, Selenium, we sing your praise,
For your alchemical touch, in awe, we raise.

THIRTY

PROFOUND VIBRATION

In the realm of elements, there lies Selenium,
A mystical force, a guardian unseen.
With an aura of light, it illuminates the night,
Guiding lost souls towards the path that's right.
    Selenium, the purifier, cleanses the impure,
Transforming darkness into a radiant allure.
Its touch, like magic, heals the wounded soul,
Restoring harmony, making broken hearts whole.
    A catalyst it is, in the alchemy of life,
Unveiling secrets, dissolving human strife.
Through its gentle presence, it brings clarity,
A beacon of wisdom, shining through obscurity.
    As the moon reflects its ethereal glow,
Selenium's essence begins to show.

A protector it stands, shielding from harm,
Embracing humanity with a loving charm.

    Oh Selenium, we sing praises to thee,
For in your presence, we find serenity.
A gift from the universe, a divine creation,
We cherish your grace, your profound vibration.

    So let us honor this element, so rare,
And bask in its aura, with utmost care.
For in Selenium's embrace, we find solace,
A sacred force, a guiding compass.

THIRTY-ONE

PURGES THE TOXINS

In the realm of elements, a guardian stands,
Selenium, the one with healing hands.
A mystic force, pure and divine,
Guiding us with its radiant shine.
    In the depths of darkness, it brings the light,
A beacon of hope, gleaming so bright.
Purifying the impurities that dwell,
Selenium's magic, a tale to tell.
    Through the alchemy of nature's grace,
It transforms, leaving no trace.
A catalyst of change, it purges the toxins,
Restoring balance, mending life's afflictions.
    Oh, Selenium, with wisdom profound,
Your presence, a treasure to be found.

A symbol of purity, so rare and true,
Our gratitude, we offer to you.
    From the earth's embrace, you emerge,
With powers that humanity can't urge.
Enlightening our path, showing the way,
Selenium, we honor you this day.
    So let us cherish this element divine,
For its mystical essence, so fine.
In gratitude, we sing your praise,
Selenium, the healer of our days.

THIRTY-TWO

CATALYTIC ABILITIES

Selenium, oh Selenium,
You are a wondrous sight.
With powers to transform,
You make everything right.
From metalloid to semiconductor,
You guide electrons with ease.
Your healing touch is felt,
By all who fall to disease.
Your beauty lies in your duality,
A protector and a source of wisdom.
You shield us from harm,
And show us the path to freedom.
With each passing moment,
You purify our souls and minds.
Your connection to the universe,
Is a treasure that we shall always find.

    Oh Selenium, we are grateful,
For the ways you have changed our lives.
Your catalytic abilities,
Are what make us thrive.
    You are a shining star,
A gift from above.
Selenium, oh Selenium,
You are the embodiment of love.

THIRTY-THREE

GUARDIAN TRUE

In the realm of elements, Selenium shines,
A mystical force, with powers divine.
A healer, a guide, it whispers in the night,
Guiding lost souls towards the path that's right.
    With gentle touch, it mends the broken heart,
Cleansing wounds, a balm for every part.
A soothing presence, it eases all pain,
A guardian of health, it shall remain.
    Through darkness it walks, a beacon of light,
A protector, a shield, in times of fright.
It guards the weary, keeps evil at bay,
Guiding them safely on their destined way.
    In its presence, shadows fade away,
Revealing truth, in every shade of gray.

Wisdom it imparts, with words so clear,
Unveiling secrets, removing all fear.

    Oh, Selenium, we honor your name,
For the gifts you bring, we shall proclaim.
A healer, a guide, a guardian true,
We're grateful for the wonders you do.

    In the realm of elements, Selenium gleams,
A force of nature, beyond our wildest dreams.
With every breath, we feel your embrace,
A symbol of hope, in this chaotic space.

THIRTY-FOUR

WE THANK YOU

In the realm of elements, Selenium resides,
A guardian of health, a healer's guide.
With atomic number thirty-four it's known,
A noble element, its virtues shown.

Selenium, a protector of utmost worth,
Defending our bodies, preserving our birth.
It shields us from harm, like armor of light,
A beacon of hope in the darkest of night.

This element of wisdom, oh so rare,
Bestows upon us knowledge, beyond compare.
Its presence enlightens, our minds it frees,
Unlocking the secrets of life's mysteries.

Selenium, a catalyst for change,
A force that transforms, in ways so strange.

It cleanses the toxins, purifies the soul,
Bringing clarity and making us whole.
  Oh Selenium, we honor your grace,
For the miracles you bring, we embrace.
In gratitude we stand, forever in awe,
For the wonders you've bestowed, without flaw.
  So let us celebrate, this element divine,
For Selenium's presence, forever we'll pine.
In every realm, in every hue,
Selenium, we thank you.

THIRTY-FIVE

HONOR SELENIUM

In the realm of elements, Selenium stands tall,
A healing touch, it carries for all.
With power and grace, it guides and protects,
A guardian of health, its effects it reflects.
    Selenium, oh Selenium, a gift from the Earth,
A mineral of value, of infinite worth.
In every cell, it weaves its magic thread,
Mending afflictions, restoring balance instead.
    With alchemical touch, it works its charm,
Unveiling clarity, with wisdom disarm.
Through the depths of darkness, it leads the way,
Illuminating truths, in the light of day.
    Selenium, oh Selenium, the alchemist's dream,
A catalyst of transformation, it does seem.

Turning the ordinary into the divine,
It elevates the spirit, with a touch so fine.
    A remedy for ailments, a balm for the soul,
Selenium's essence makes us whole.
With every breath, its potency we feel,
A guardian of health, its presence is real.
    So let us honor Selenium, in all its might,
Embracing its power, in healing's pure light.
For in the realm of elements, it shines bright,
Guiding us towards wellness, with love and insight.

ABOUT THE AUTHOR

Walter the Educator is one of the pseudonyms for Walter Anderson. Formally educated in Chemistry, Business, and Education, he is an educator, an author, a diverse entrepreneur, and he is the son of a disabled war veteran. "Walter the Educator" shares his time between educating and creating. He holds interests and owns several creative projects that entertain, enlighten, enhance, and educate, hoping to inspire and motivate you.

Follow, find new works, and stay up to date
with Walter the Educator™
at WaltertheEducator.com

www.ingramcontent.com/pod-product-compliance
Lightning Source LLC
LaVergne TN
LVHW051959060526
838201LV00059B/3743